Company De

Notes

Date	Appointed Calibrator	Signature

Index

Equipment to be Calibrated	Ref No.	Page

Index

Equipment to be Calibrated	Ref No.	Page

Index

Equipment to be Calibrated	Ref No.	Page

Index

Equipment to be Calibrated	Ref No.	Page

Item Name	Identifier Ref
Supplier	**Date Purchased**

Calibration requirements and Inspection Intervals

Date	Initials	Notes	Days till Next Calibration	Next Calibration Date

Continuation Sheet

Item Name________________________ Ref:______________

Date	Initials	Notes	Days till Next Calibration	Next Calibration Date

Item Name	Identifier Ref
Supplier	**Date Purchased**

Calibration requirements and Inspection Intervals

Date	Initials	Notes	Days till Next Calibration	Next Calibration Date

Continuation Sheet

Item Name________________________________ Ref:__________________

Date	Initials	Notes	Days till Next Calibration	Next Calibration Date

Item Name	Identifier Ref
Supplier	**Date Purchased**

Calibration requirements and Inspection Intervals

Date	Initials	Notes	Days till Next Calibration	Next Calibration Date

Continuation Sheet

Item Name____________________________ Ref:________________

Date	Initials	Notes	Days till Next Calibration	Next Calibration Date

Item Name	Identifier Ref
Supplier	Date Purchased

Calibration requirements and Inspection Intervals

Date	Initials	Notes	Days till Next Calibration	Next Calibration Date

Continuation Sheet

Item Name________________________________ Ref:__________________

Date	Initials	Notes	Days till Next Calibration	Next Calibration Date

Item Name	Identifier Ref
Supplier	**Date Purchased**

Calibration requirements and Inspection Intervals

Date	Initials	Notes	Days till Next Calibration	Next Calibration Date

Continuation Sheet

Item Name______________________________ Ref:__________________

Date	Initials	Notes	Days till Next Calibration	Next Calibration Date

Item Name	Identifier Ref

Supplier	Date Purchased

Calibration requirements and Inspection Intervals

Date	Initials	Notes	Days till Next Calibration	Next Calibration Date

Continuation Sheet

Item Name________________________ Ref:____________

Date	Initials	Notes	Days till Next Calibration	Next Calibration Date

Item Name	Identifier Ref

Supplier	Date Purchased

Calibration requirements and Inspection Intervals

Date	Initials	Notes	Days till Next Calibration	Next Calibration Date

Continuation Sheet

Item Name______________________________ Ref:________________

Date	Initials	Notes	Days till Next Calibration	Next Calibration Date

Item Name	Identifier Ref

Supplier	Date Purchased

Calibration requirements and Inspection Intervals

Date	Initials	Notes	Days till Next Calibration	Next Calibration Date

Continuation Sheet

Item Name________________________________ Ref:__________________

Date	Initials	Notes	Days till Next Calibration	Next Calibration Date

Item Name	Identifier Ref

Supplier	Date Purchased

Calibration requirements and Inspection Intervals

Date	Initials	Notes	Days till Next Calibration	Next Calibration Date

Continuation Sheet

Item Name______________________________ Ref:________________

Date	Initials	Notes	Days till Next Calibration	Next Calibration Date

Item Name	Identifier Ref

Supplier	Date Purchased

Calibration requirements and Inspection Intervals

Date	Initials	Notes	Days till Next Calibration	Next Calibration Date

Continuation Sheet

Item Name______________________________ Ref:_________________

Date	Initials	Notes	Days till Next Calibration	Next Calibration Date

Item Name	Identifier Ref

Supplier	Date Purchased

Calibration requirements and Inspection Intervals

Date	Initials	Notes	Days till Next Calibration	Next Calibration Date

Continuation Sheet

Item Name______________________________ Ref:________________

Date	Initials	Notes	Days till Next Calibration	Next Calibration Date

Item Name	Identifier Ref

Supplier	Date Purchased

Calibration requirements and Inspection Intervals

Date	Initials	Notes	Days till Next Calibration	Next Calibration Date

Continuation Sheet

Item Name______________________________ Ref:________________

Date	Initials	Notes	Days till Next Calibration	Next Calibration Date

Item Name	Identifier Ref
Supplier	**Date Purchased**

Calibration requirements and Inspection Intervals

Date	Initials	Notes	Days till Next Calibration	Next Calibration Date

Continuation Sheet

Item Name____________________________ Ref:________________

Date	Initials	Notes	Days till Next Calibration	Next Calibration Date

Item Name	Identifier Ref
Supplier	**Date Purchased**

Calibration requirements and Inspection Intervals

Date	Initials	Notes	Days till Next Calibration	Next Calibration Date

Continuation Sheet

Item Name______________________________ Ref:_________________

Date	Initials	Notes	Days till Next Calibration	Next Calibration Date

Item Name	Identifier Ref

Supplier	Date Purchased

Calibration requirements and Inspection Intervals

Date	Initials	Notes	Days till Next Calibration	Next Calibration Date

Continuation Sheet

Item Name________________________ Ref:______________

Date	Initials	Notes	Days till Next Calibration	Next Calibration Date

Item Name	Identifier Ref
Supplier	**Date Purchased**

Calibration requirements and Inspection Intervals

Date	Initials	Notes	Days till Next Calibration	Next Calibration Date

Continuation Sheet

Item Name________________________ Ref:______________

Date	Initials	Notes	Days till Next Calibration	Next Calibration Date

Item Name	Identifier Ref
Supplier	**Date Purchased**

Calibration requirements and Inspection Intervals

Date	Initials	Notes	Days till Next Calibration	Next Calibration Date

Continuation Sheet

Item Name______________________ Ref:____________

Date	Initials	Notes	Days till Next Calibration	Next Calibration Date

Item Name	Identifier Ref

Supplier	Date Purchased

Calibration requirements and Inspection Intervals

Date	Initials	Notes	Days till Next Calibration	Next Calibration Date

Continuation Sheet

Item Name________________________ Ref:______________

Date	Initials	Notes	Days till Next Calibration	Next Calibration Date

Item Name	Identifier Ref
Supplier	**Date Purchased**

Calibration requirements and Inspection Intervals

Date	Initials	Notes	Days till Next Calibration	Next Calibration Date

Continuation Sheet

Item Name________________________ Ref:______________

Date	Initials	Notes	Days till Next Calibration	Next Calibration Date

Item Name	Identifier Ref
Supplier	**Date Purchased**

Calibration requirements and Inspection Intervals

Date	Initials	Notes	Days till Next Calibration	Next Calibration Date

Continuation Sheet

Item Name______________________________ Ref:________________

Date	Initials	Notes	Days till Next Calibration	Next Calibration Date

Item Name	Identifier Ref
Supplier	**Date Purchased**

Calibration requirements and Inspection Intervals

Date	Initials	Notes	Days till Next Calibration	Next Calibration Date

Continuation Sheet

Item Name______________________________ Ref:________________

Date	Initials	Notes	Days till Next Calibration	Next Calibration Date

Item Name	Identifier Ref
Supplier	**Date Purchased**

Calibration requirements and Inspection Intervals

Date	Initials	Notes	Days till Next Calibration	Next Calibration Date

Continuation Sheet

Item Name______________________________ Ref:_________________

Date	Initials	Notes	Days till Next Calibration	Next Calibration Date

Item Name	Identifier Ref
Supplier	**Date Purchased**

Calibration requirements and Inspection Intervals

Date	Initials	Notes	Days till Next Calibration	Next Calibration Date

Continuation Sheet

Item Name________________________________ Ref:__________________

Date	Initials	Notes	Days till Next Calibration	Next Calibration Date

Item Name	Identifier Ref

Supplier	Date Purchased

Calibration requirements and Inspection Intervals

Date	Initials	Notes	Days till Next Calibration	Next Calibration Date

Continuation Sheet

Item Name______________________________ Ref:_______________

Date	Initials	Notes	Days till Next Calibration	Next Calibration Date

Item Name	Identifier Ref

Supplier	Date Purchased

Calibration requirements and Inspection Intervals

Date	Initials	Notes	Days till Next Calibration	Next Calibration Date

Continuation Sheet

Item Name______________________________ Ref:_________________

Date	Initials	Notes	Days till Next Calibration	Next Calibration Date

Item Name	Identifier Ref

Supplier	Date Purchased

Calibration requirements and Inspection Intervals

Date	Initials	Notes	Days till Next Calibration	Next Calibration Date

Continuation Sheet

Item Name________________________ Ref:______________

Date	Initials	Notes	Days till Next Calibration	Next Calibration Date

Item Name	Identifier Ref
Supplier	**Date Purchased**

Calibration requirements and Inspection Intervals

Date	Initials	Notes	Days till Next Calibration	Next Calibration Date

Continuation Sheet

Item Name______________________________ Ref:_________________

Date	Initials	Notes	Days till Next Calibration	Next Calibration Date

Item Name	Identifier Ref
Supplier	**Date Purchased**

Calibration requirements and Inspection Intervals

Date	Initials	Notes	Days till Next Calibration	Next Calibration Date

Continuation Sheet

Item Name________________________ Ref:____________

Date	Initials	Notes	Days till Next Calibration	Next Calibration Date

Item Name	Identifier Ref

Supplier	Date Purchased

Calibration requirements and Inspection Intervals

Date	Initials	Notes	Days till Next Calibration	Next Calibration Date

Continuation Sheet

Item Name____________________________ Ref:________________

Date	Initials	Notes	Days till Next Calibration	Next Calibration Date

Item Name	Identifier Ref
Supplier	**Date Purchased**

Calibration requirements and Inspection Intervals

Date	Initials	Notes	Days till Next Calibration	Next Calibration Date

Continuation Sheet

Item Name______________________________ Ref:________________

Date	Initials	Notes	Days till Next Calibration	Next Calibration Date

Item Name	Identifier Ref
Supplier	**Date Purchased**

Calibration requirements and Inspection Intervals

Date	Initials	Notes	Days till Next Calibration	Next Calibration Date

Continuation Sheet

Item Name________________________ Ref:______________

Date	Initials	Notes	Days till Next Calibration	Next Calibration Date

Item Name	Identifier Ref

Supplier	Date Purchased

Calibration requirements and Inspection Intervals

Date	Initials	Notes	Days till Next Calibration	Next Calibration Date

Continuation Sheet

Item Name______________________________ Ref:________________

Date	Initials	Notes	Days till Next Calibration	Next Calibration Date

Item Name	Identifier Ref

Supplier	Date Purchased

Calibration requirements and Inspection Intervals

Date	Initials	Notes	Days till Next Calibration	Next Calibration Date

Continuation Sheet

Item Name________________________ Ref:____________

Date	Initials	Notes	Days till Next Calibration	Next Calibration Date

Item Name	Identifier Ref

Supplier	Date Purchased

Calibration requirements and Inspection Intervals

Date	Initials	Notes	Days till Next Calibration	Next Calibration Date

Continuation Sheet

Item Name__________________________ Ref:_______________

Date	Initials	Notes	Days till Next Calibration	Next Calibration Date

Item Name	Identifier Ref
Supplier	**Date Purchased**

Calibration requirements and Inspection Intervals

Date	Initials	Notes	Days till Next Calibration	Next Calibration Date

Continuation Sheet

Item Name____________________________ Ref:________________

Date	Initials	Notes	Days till Next Calibration	Next Calibration Date

Item Name	Identifier Ref
Supplier	Date Purchased
Calibration requirements and Inspection Intervals	

Date	Initials	Notes	Days till Next Calibration	Next Calibration Date

Continuation Sheet

Item Name________________________________ Ref:__________________

Date	Initials	Notes	Days till Next Calibration	Next Calibration Date

Item Name	Identifier Ref
Supplier	**Date Purchased**

Calibration requirements and Inspection Intervals

Date	Initials	Notes	Days till Next Calibration	Next Calibration Date

Continuation Sheet

Item Name________________________ Ref:______________

Date	Initials	Notes	Days till Next Calibration	Next Calibration Date

Item Name	Identifier Ref

Supplier	Date Purchased

Calibration requirements and Inspection Intervals

Date	Initials	Notes	Days till Next Calibration	Next Calibration Date

Continuation Sheet

Item Name______________________________ Ref:_________________

Date	Initials	Notes	Days till Next Calibration	Next Calibration Date

Item Name	Identifier Ref
Supplier	**Date Purchased**
Calibration requirements and Inspection Intervals	

Date	Initials	Notes	Days till Next Calibration	Next Calibration Date

Continuation Sheet

Item Name______________________________ Ref:________________

Date	Initials	Notes	Days till Next Calibration	Next Calibration Date

Item Name	Identifier Ref

Supplier	Date Purchased

Calibration requirements and Inspection Intervals

Date	Initials	Notes	Days till Next Calibration	Next Calibration Date

Continuation Sheet

Item Name________________________ Ref:______________

Date	Initials	Notes	Days till Next Calibration	Next Calibration Date

Item Name	Identifier Ref

Supplier	Date Purchased

Calibration requirements and Inspection Intervals

Date	Initials	Notes	Days till Next Calibration	Next Calibration Date

Continuation Sheet

Item Name____________________________ Ref:________________

Date	Initials	Notes	Days till Next Calibration	Next Calibration Date

Item Name	Identifier Ref

Supplier	Date Purchased

Calibration requirements and Inspection Intervals

Date	Initials	Notes	Days till Next Calibration	Next Calibration Date

Continuation Sheet

Item Name______________________ Ref:____________

Date	Initials	Notes	Days till Next Calibration	Next Calibration Date

Item Name	Identifier Ref
Supplier	**Date Purchased**

Calibration requirements and Inspection Intervals

Date	Initials	Notes	Days till Next Calibration	Next Calibration Date

Continuation Sheet

Item Name________________________________ Ref:__________________

Date	Initials	Notes	Days till Next Calibration	Next Calibration Date

Item Name	Identifier Ref

Supplier	Date Purchased

Calibration requirements and Inspection Intervals

Date	Initials	Notes	Days till Next Calibration	Next Calibration Date

Continuation Sheet

Item Name________________________ Ref:______________

Date	Initials	Notes	Days till Next Calibration	Next Calibration Date

Item Name	Identifier Ref
Supplier	**Date Purchased**

Calibration requirements and Inspection Intervals

Date	Initials	Notes	Days till Next Calibration	Next Calibration Date

Continuation Sheet

Item Name____________________________ Ref:________________

Date	Initials	Notes	Days till Next Calibration	Next Calibration Date

Item Name	Identifier Ref

Supplier	Date Purchased

Calibration requirements and Inspection Intervals

Date	Initials	Notes	Days till Next Calibration	Next Calibration Date

Continuation Sheet

Item Name______________________________ Ref:________________

Date	Initials	Notes	Days till Next Calibration	Next Calibration Date

Item Name	Identifier Ref
Supplier	**Date Purchased**

Calibration requirements and Inspection Intervals

Date	Initials	Notes	Days till Next Calibration	Next Calibration Date

Continuation Sheet

Item Name______________________________ Ref:________________

Date	Initials	Notes	Days till Next Calibration	Next Calibration Date

Item Name	Identifier Ref

Supplier	Date Purchased

Calibration requirements and Inspection Intervals

Date	Initials	Notes	Days till Next Calibration	Next Calibration Date

Continuation Sheet

Item Name________________________ Ref:______________

Date	Initials	Notes	Days till Next Calibration	Next Calibration Date

Item Name	Identifier Ref
Supplier	**Date Purchased**

Calibration requirements and Inspection Intervals

Date	Initials	Notes	Days till Next Calibration	Next Calibration Date

Continuation Sheet

Item Name________________________________ Ref:__________________

Date	Initials	Notes	Days till Next Calibration	Next Calibration Date

Item Name	Identifier Ref
Supplier	**Date Purchased**

Calibration requirements and Inspection Intervals

Date	Initials	Notes	Days till Next Calibration	Next Calibration Date

Continuation Sheet

Item Name______________________________ Ref:_________________

Date	Initials	Notes	Days till Next Calibration	Next Calibration Date

Item Name	Identifier Ref

Supplier	Date Purchased

Calibration requirements and Inspection Intervals

Date	Initials	Notes	Days till Next Calibration	Next Calibration Date

Continuation Sheet

Item Name______________________ Ref:____________

Date	Initials	Notes	Days till Next Calibration	Next Calibration Date

Item Name	Identifier Ref

Supplier	Date Purchased

Calibration requirements and Inspection Intervals

Date	Initials	Notes	Days till Next Calibration	Next Calibration Date

Continuation Sheet

Item Name______________________________ Ref:_________________

Date	Initials	Notes	Days till Next Calibration	Next Calibration Date

Item Name	Identifier Ref

Supplier	Date Purchased

Calibration requirements and Inspection Intervals

Date	Initials	Notes	Days till Next Calibration	Next Calibration Date

Continuation Sheet

Item Name________________________ Ref:______________

Date	Initials	Notes	Days till Next Calibration	Next Calibration Date

Item Name	Identifier Ref

Supplier	Date Purchased

Calibration requirements and Inspection Intervals

Date	Initials	Notes	Days till Next Calibration	Next Calibration Date

Continuation Sheet

Item Name________________________________ Ref:_______________

Date	Initials	Notes	Days till Next Calibration	Next Calibration Date

Item Name	Identifier Ref
Supplier	**Date Purchased**

Calibration requirements and Inspection Intervals

Date	Initials	Notes	Days till Next Calibration	Next Calibration Date

Continuation Sheet

Item Name______________________________ Ref:________________

Date	Initials	Notes	Days till Next Calibration	Next Calibration Date

Item Name	Identifier Ref

Supplier	Date Purchased

Calibration requirements and Inspection Intervals

Date	Initials	Notes	Days till Next Calibration	Next Calibration Date

Continuation Sheet

Item Name________________________ Ref:______________

Date	Initials	Notes	Days till Next Calibration	Next Calibration Date

Item Name	Identifier Ref
Supplier	**Date Purchased**

Calibration requirements and Inspection Intervals

Date	Initials	Notes	Days till Next Calibration	Next Calibration Date

Continuation Sheet

Item Name______________________________ Ref:________________

Date	Initials	Notes	Days till Next Calibration	Next Calibration Date

Item Name	Identifier Ref

Supplier	Date Purchased

Calibration requirements and Inspection Intervals

Date	Initials	Notes	Days till Next Calibration	Next Calibration Date

Continuation Sheet

Item Name______________________ Ref:____________

Date	Initials	Notes	Days till Next Calibration	Next Calibration Date

Item Name	Identifier Ref
Supplier	**Date Purchased**

Calibration requirements and Inspection Intervals

Date	Initials	Notes	Days till Next Calibration	Next Calibration Date

Continuation Sheet

Item Name________________________ Ref:______________

Date	Initials	Notes	Days till Next Calibration	Next Calibration Date

Item Name	Identifier Ref

Supplier	Date Purchased

Calibration requirements and Inspection Intervals

Date	Initials	Notes	Days till Next Calibration	Next Calibration Date

Continuation Sheet

Item Name______________________ Ref:______________

Date	Initials	Notes	Days till Next Calibration	Next Calibration Date

Item Name	Identifier Ref
Supplier	**Date Purchased**

Calibration requirements and Inspection Intervals

Date	Initials	Notes	Days till Next Calibration	Next Calibration Date

Continuation Sheet

Item Name______________________________ Ref:_________________

Date	Initials	Notes	Days till Next Calibration	Next Calibration Date

Item Name	Identifier Ref

Supplier	Date Purchased

Calibration requirements and Inspection Intervals

Date	Initials	Notes	Days till Next Calibration	Next Calibration Date

Continuation Sheet

Item Name________________________________ Ref:__________________

Date	Initials	Notes	Days till Next Calibration	Next Calibration Date

Item Name	Identifier Ref

Supplier	Date Purchased

Calibration requirements and Inspection Intervals

Date	Initials	Notes	Days till Next Calibration	Next Calibration Date

Continuation Sheet

Item Name______________________________ Ref:_________________

Date	Initials	Notes	Days till Next Calibration	Next Calibration Date

Item Name	Identifier Ref
Supplier	**Date Purchased**

Calibration requirements and Inspection Intervals

Date	Initials	Notes	Days till Next Calibration	Next Calibration Date

Continuation Sheet

Item Name______________________________ Ref:________________

Date	Initials	Notes	Days till Next Calibration	Next Calibration Date

Item Name	Identifier Ref
Supplier	**Date Purchased**

Calibration requirements and Inspection Intervals

Date	Initials	Notes	Days till Next Calibration	Next Calibration Date

Continuation Sheet

Item Name______________________________ Ref:_________________

Date	Initials	Notes	Days till Next Calibration	Next Calibration Date

Item Name	Identifier Ref

Supplier	Date Purchased

Calibration requirements and Inspection Intervals

Date	Initials	Notes	Days till Next Calibration	Next Calibration Date

Continuation Sheet

Item Name______________________________ Ref:________________

Date	Initials	Notes	Days till Next Calibration	Next Calibration Date

Item Name	Identifier Ref

Supplier	Date Purchased

Calibration requirements and Inspection Intervals

Date	Initials	Notes	Days till Next Calibration	Next Calibration Date

Continuation Sheet

Item Name______________________________ Ref:_________________

Date	Initials	Notes	Days till Next Calibration	Next Calibration Date

Item Name	Identifier Ref
Supplier	**Date Purchased**

Calibration requirements and Inspection Intervals

Date	Initials	Notes	Days till Next Calibration	Next Calibration Date

Continuation Sheet

Item Name______________________________ Ref:________________

Date	Initials	Notes	Days till Next Calibration	Next Calibration Date

Item Name	Identifier Ref
Supplier	**Date Purchased**

Calibration requirements and Inspection Intervals

Date	Initials	Notes	Days till Next Calibration	Next Calibration Date

Continuation Sheet

Item Name______________________ Ref:____________

Date	Initials	Notes	Days till Next Calibration	Next Calibration Date

Item Name	Identifier Ref

Supplier	Date Purchased

Calibration requirements and Inspection Intervals

Date	Initials	Notes	Days till Next Calibration	Next Calibration Date

Continuation Sheet

Item Name________________________________ Ref:________________

Date	Initials	Notes	Days till Next Calibration	Next Calibration Date

Item Name	Identifier Ref

Supplier	Date Purchased

Calibration requirements and Inspection Intervals

Date	Initials	Notes	Days till Next Calibration	Next Calibration Date

Continuation Sheet

Item Name____________________________ Ref:________________

Date	Initials	Notes	Days till Next Calibration	Next Calibration Date

Item Name	Identifier Ref
Supplier	**Date Purchased**

Calibration requirements and Inspection Intervals

Date	Initials	Notes	Days till Next Calibration	Next Calibration Date

Continuation Sheet

Item Name______________________________ Ref:________________

Date	Initials	Notes	Days till Next Calibration	Next Calibration Date

Item Name	Identifier Ref

Supplier	Date Purchased

Calibration requirements and Inspection Intervals

Date	Initials	Notes	Days till Next Calibration	Next Calibration Date

Continuation Sheet

Item Name______________________ Ref:____________

Date	Initials	Notes	Days till Next Calibration	Next Calibration Date

Item Name	Identifier Ref

Supplier	Date Purchased

Calibration requirements and Inspection Intervals

Date	Initials	Notes	Days till Next Calibration	Next Calibration Date

Continuation Sheet

Item Name____________________________ Ref:________________

Date	Initials	Notes	Days till Next Calibration	Next Calibration Date

Item Name	Identifier Ref
Supplier	**Date Purchased**

Calibration requirements and Inspection Intervals

Date	Initials	Notes	Days till Next Calibration	Next Calibration Date

Continuation Sheet

Item Name________________________ Ref:______________

Date	Initials	Notes	Days till Next Calibration	Next Calibration Date

Item Name	Identifier Ref

Supplier	Date Purchased

Calibration requirements and Inspection Intervals

Date	Initials	Notes	Days till Next Calibration	Next Calibration Date

Continuation Sheet

Item Name______________________________ Ref:________________

Date	Initials	Notes	Days till Next Calibration	Next Calibration Date

Item Name	Identifier Ref
Supplier	**Date Purchased**

Calibration requirements and Inspection Intervals

Date	Initials	Notes	Days till Next Calibration	Next Calibration Date

Continuation Sheet

Item Name______________________________ Ref:_________________

Date	Initials	Notes	Days till Next Calibration	Next Calibration Date

Item Name	Identifier Ref

Supplier	Date Purchased

Calibration requirements and Inspection Intervals

Date	Initials	Notes	Days till Next Calibration	Next Calibration Date

Continuation Sheet

Item Name______________________________ Ref:________________

Date	Initials	Notes	Days till Next Calibration	Next Calibration Date

Item Name	Identifier Ref

Supplier	Date Purchased

Calibration requirements and Inspection Intervals

Date	Initials	Notes	Days till Next Calibration	Next Calibration Date

Continuation Sheet

Item Name______________________________ Ref:________________

Date	Initials	Notes	Days till Next Calibration	Next Calibration Date

Item Name	Identifier Ref

Supplier	Date Purchased

Calibration requirements and Inspection Intervals

Date	Initials	Notes	Days till Next Calibration	Next Calibration Date

Continuation Sheet

Item Name______________________ Ref:____________

Date	Initials	Notes	Days till Next Calibration	Next Calibration Date

Item Name	Identifier Ref
Supplier	Date Purchased

Calibration requirements and Inspection Intervals

Date	Initials	Notes	Days till Next Calibration	Next Calibration Date

Continuation Sheet

Item Name______________________________ Ref:_________________

Date	Initials	Notes	Days till Next Calibration	Next Calibration Date

Item Name	Identifier Ref

Supplier	Date Purchased

Calibration requirements and Inspection Intervals

Date	Initials	Notes	Days till Next Calibration	Next Calibration Date

Continuation Sheet

Item Name______________________ Ref:____________

Date	Initials	Notes	Days till Next Calibration	Next Calibration Date

Item Name	Identifier Ref

Supplier	Date Purchased

Calibration requirements and Inspection Intervals

Date	Initials	Notes	Days till Next Calibration	Next Calibration Date

Continuation Sheet

Item Name________________________ Ref:______________

Date	Initials	Notes	Days till Next Calibration	Next Calibration Date

Item Name	Identifier Ref

Supplier	Date Purchased

Calibration requirements and Inspection Intervals

Date	Initials	Notes	Days till Next Calibration	Next Calibration Date

Continuation Sheet

Item Name______________________________ Ref:________________

Date	Initials	Notes	Days till Next Calibration	Next Calibration Date

Item Name	Identifier Ref

Supplier	Date Purchased

Calibration requirements and Inspection Intervals

Date	Initials	Notes	Days till Next Calibration	Next Calibration Date

Continuation Sheet

Item Name______________________ Ref:____________

Date	Initials	Notes	Days till Next Calibration	Next Calibration Date

Item Name	Identifier Ref

Supplier	Date Purchased

Calibration requirements and Inspection Intervals

Date	Initials	Notes	Days till Next Calibration	Next Calibration Date

Continuation Sheet

Item Name______________________________ Ref:________________

Date	Initials	Notes	Days till Next Calibration	Next Calibration Date

Item Name	Identifier Ref
Supplier	**Date Purchased**

Calibration requirements and Inspection Intervals

Date	Initials	Notes	Days till Next Calibration	Next Calibration Date

Continuation Sheet

Item Name______________________________ Ref:_________________

Date	Initials	Notes	Days till Next Calibration	Next Calibration Date

Item Name	Identifier Ref

Supplier	Date Purchased

Calibration requirements and Inspection Intervals

Date	Initials	Notes	Days till Next Calibration	Next Calibration Date

Continuation Sheet

Item Name______________________________ Ref:_________________

Date	Initials	Notes	Days till Next Calibration	Next Calibration Date

Item Name	Identifier Ref
Supplier	**Date Purchased**

Calibration requirements and Inspection Intervals

Date	Initials	Notes	Days till Next Calibration	Next Calibration Date

Continuation Sheet

Item Name________________________ Ref:______________

Date	Initials	Notes	Days till Next Calibration	Next Calibration Date

Item Name	Identifier Ref

Supplier	Date Purchased

Calibration requirements and Inspection Intervals

Date	Initials	Notes	Days till Next Calibration	Next Calibration Date

Continuation Sheet

Item Name______________________ Ref:____________

Date	Initials	Notes	Days till Next Calibration	Next Calibration Date

Item Name	Identifier Ref
Supplier	**Date Purchased**

Calibration requirements and Inspection Intervals

Date	Initials	Notes	Days till Next Calibration	Next Calibration Date

Continuation Sheet

Item Name______________________________ Ref:________________

Date	Initials	Notes	Days till Next Calibration	Next Calibration Date

Item Name	Identifier Ref

Supplier	Date Purchased

Calibration requirements and Inspection Intervals

Date	Initials	Notes	Days till Next Calibration	Next Calibration Date

Continuation Sheet

Item Name______________________________ Ref:_______________

Date	Initials	Notes	Days till Next Calibration	Next Calibration Date

Item Name	Identifier Ref
Supplier	**Date Purchased**

Calibration requirements and Inspection Intervals

Date	Initials	Notes	Days till Next Calibration	Next Calibration Date

Continuation Sheet

Item Name________________________________ Ref:__________________

Date	Initials	Notes	Days till Next Calibration	Next Calibration Date

Item Name	Identifier Ref
Supplier	**Date Purchased**

Calibration requirements and Inspection Intervals

Date	Initials	Notes	Days till Next Calibration	Next Calibration Date

Continuation Sheet

Item Name______________________ Ref:____________

Date	Initials	Notes	Days till Next Calibration	Next Calibration Date

Item Name	Identifier Ref
Supplier	**Date Purchased**

Calibration requirements and Inspection Intervals

Date	Initials	Notes	Days till Next Calibration	Next Calibration Date

Continuation Sheet

Item Name______________________________ Ref:_________________

Date	Initials	Notes	Days till Next Calibration	Next Calibration Date

Item Name	Identifier Ref

Supplier	Date Purchased

Calibration requirements and Inspection Intervals

Date	Initials	Notes	Days till Next Calibration	Next Calibration Date

Continuation Sheet

Item Name________________________ Ref:______________

Date	Initials	Notes	Days till Next Calibration	Next Calibration Date

Item Name	Identifier Ref

Supplier	Date Purchased

Calibration requirements and Inspection Intervals

Date	Initials	Notes	Days till Next Calibration	Next Calibration Date

Continuation Sheet

Item Name______________________________ Ref:________________

Date	Initials	Notes	Days till Next Calibration	Next Calibration Date

Item Name	Identifier Ref
Supplier	Date Purchased

Calibration requirements and Inspection Intervals

Date	Initials	Notes	Days till Next Calibration	Next Calibration Date

Continuation Sheet

Item Name_______________________________ Ref:_________________

Date	Initials	Notes	Days till Next Calibration	Next Calibration Date

Printed in Great Britain
by Amazon